How Do You Live There?

LIVING IN THE RAIN FOREST

Alicia Z. Klepeis

PowerKiDS press

NEW YORK

Published in 2021 by The Rosen Publishing Group, Inc.
29 East 21st Street, New York, NY 10010

First edition

Editor: Kristen Susienka
Designer: Rachel Rising

Photo Credits: Cover, p. 1 Kunlaphat Raksakul/Shutterstock.com; Cover, pp. 3,4,6,7,8,9,10,11,12,13,14,16,18,20,21,22,24,26,28,29,30,31,32 (background) Yevhenii Borshosh/Shutterstock.com; p. 5 ronnybas frimages/Shutterstock.com; p. 7 Edmund Lowe Photography/Shutterstock.com; p. 8 Dirk Ercken/Shutterstock.com; p. 9 Chokniti Khongchum/Shutterstock.com; p. 10 Imagentle/Shutterstock.com; p. 11 Gabriele Maltinti/Shutterstock.com; p. 13 Mazur Travel/Shutterstock.com; p. 15 Muslianshah Masrie/Shutterstock.com; p. 17 Anna ART/Shutterstock.com; p. 19 Daniel Andis/Shutterstock.com; p. 20 marchello74/Shutterstock.com; p. 21 Said Marroun/Shutterstock.com; p. 23 Pikoso.kz/Shutterstock.com; p. 25 Ivan Kuzmin/Shutterstock.com; p. 27 Rich Carey/Shutterstock.com; p. 28 Sara Berdon/Shutterstock.com; p. 29 Eco Images/Universal Images Group/Getty Images; p. 30 Richard Whitcombe/Shutterstock.com.

Library of Congress Cataloging-in-Publication Data

Names: Klepeis, Alicia, 1971- author.
Title: Living in the rain forest / Alicia Z. Klepeis.
Description: New York : PowerKids Press, [2021] | Series: How do you live there?! | Includes index.
Identifiers: LCCN 2019051209 | ISBN 9781725316553 (paperback) | ISBN 9781725316577 (library binding) | ISBN 9781725316560 (6 pack)
Subjects: LCSH: Rain forest ecology--Juvenile literature.
Classification: LCC QH541.5.R27 K59 2021 | DDC 577.34--dc23
LC record available at https://lccn.loc.gov/2019051209

Manufactured in the United States of America

CPSIA Compliance Information: Batch #CSPK20. For further information contact Rosen Publishing, New York, New York at 1-800-237-9932.

CONTENTS

WHAT IS THE RAIN FOREST?

Rain forests are areas with tall, mainly **evergreen** trees and lots of rainfall. Many animals, plants, and people live in rain forests. About 6 percent of Earth's land is rain forest. Around half of all plant and animal **species** on Earth live there.

There are different types of rain forest. Two common ones are **tropical** and **temperate**. Most rain forests are tropical. Their climate is warm and moist. It rains a lot! Tropical rain forests are usually found near the **equator**. Australia has tropical rain forests. So does the island of New Guinea. Asia's tropical rain forests are in Southeast Asia and western India. Central and South America have rain forests too. The Congo Basin in Africa is a tropical rain forest region that stretches across six countries!

The world's temperate rain forests are located along coasts, farther from the equator. Weather there is less hot and humid than in tropical rain forests, but it rains a lot there too.

Some of the biggest temperate rain forests are on North America's Pacific coast. They stretch between the U.S. states of California and Alaska, passing through British Columbia in Canada. They contain giant redwoods, the world's tallest trees. In Europe, temperate rain forests are found in the United Kingdom and Norway. South America has them along the coast of Chile. In Asia, Japan has temperate rain forests too. Both southern Australia and New Zealand have these rain forests. Rain forests might look different depending on where they are located, but they all have similar features.

A small stream winds its way among mossy rocks in this Pacific Northwest rain forest.

So Much Rain!

All rain forests get a lot of rain. Tropical rain forests get about 80 to 400 inches (200 to 1,000 centimeters) each year. The air in these rain forests is warm and wet all year long. Temperatures don't change much between night and day. Temperate rain forests are cooler than tropical ones. They get about 60 to 200 inches (150 to 500 cm) of rain per year.

LAYERS OF A RAIN FOREST

Rain forests have four layers, or levels: the emergent layer, the canopy, the understory, and the forest floor. Each layer has its own plants and animals living in it. Layers are not all the same in every rain forest, however. They start and end at different heights depending on the forest.

The emergent layer is the highest part of the forest. Treetops wave in the sky. Eagles and other birds build nests in the branches. The canopy is crowded with leaves and branches. It is home to monkeys, bats, and smaller birds.

TREE FROG

The rain forest of Thailand's Khao Yai National Park has many trees growing close together.

The understory has small bushes, short trees, and some **rare** animals. Jaguars, snakes, and tree frogs roam in the understory. The forest floor is littered with tree roots, dirt, and plants that don't need a lot of sunlight.

CATCHING SUNLIGHT

The different layers of the rain forest receive different amounts of sunlight. The sunlight is strongest at the top of the forest. Farther down the layers, sunlight becomes weaker. The canopy is thick with plants, blocking out most of the sunlight for the layers below.

The plants in different layers have different features that help them adapt to the amount of sunlight they receive. Many plants in the understory have big leaves so they can take in as much sunlight as possible. Flowers called orchids have

Sunlight streams through the trees in a rain forest on the Caribbean island of Guadeloupe.

adapted in a different way. Instead of growing in the ground, many orchids can grow on tree branches, with their roots in the air. By growing high up in the trees, orchids get more sunlight than they would on the forest floor.

AMAZING ANIMALS AND PLANTS

Most kinds of plants on Earth grow in tropical rain forests. Trees, ferns, and mosses are just a few examples. Rain forests also have lots of weird plants. One is the world's largest flower, the rafflesia. It can grow to be more than 3 feet (1 meter) across! It smells like rotting meat, which is gross to humans but tasty to flies. When the flies land on the flower, they help it create new seeds.

More species of animals live in tropical rain forests than anywhere else on Earth. Orangutans live in the rain forests on the Indonesian islands of Borneo and Sumatra. Lemurs swing from branches in the rain forests of Madagascar. Colorful strawberry poison dart frogs live in Central American rain forests. And capybaras, the world's largest rodents, live in South American rain forests.

Rafflesia flowers, like this one in Indonesia, can weigh up to 15 pounds (6.8 kilograms).

The Amazon Rain Forest

The largest tropical rain forest on Earth is the Amazon Rain Forest in South America. If it were a country, it would be the ninth biggest in the world! Colorful birds like the toco toucan and scarlet macaw, jaguars, and bright blue morpho butterflies all make their homes in the Amazon. In the rain forest's Amazon River, there are rare pink dolphins, otters, and flesh-eating fish called piranhas. About 2.5 million different insects live here too.

LIVING IN THE RAIN FOREST

Animals and plants aren't the only things living in rain forests. People live in rain forests too. However, living in the rain forest is tough. The soil is not great for growing food because there is so much rain. Trees and plants have to be removed to build homes. Many dangerous insects and animals live in rain forests. People who live there have to be careful.

Many native groups live in rain forests. The Yanomami and Kayapó **tribes** live in the Amazon Rain Forest. The Baka and Efe live in Africa's Congo Basin. These and other native peoples figured out how to survive in the rain forest's tough environment. They built homes and communities, and made tools to hunt with. They have lived in the area for many **generations**.

A man from a tribe called the Mentawai poses for a picture. He and his tribe live in the forests on islands in Indonesia.

Traditionally, native peoples hunted, fished, and gathered their foods. It was common to burn down trees and plants to make room for farming.

Today, many people who live in the rain forest still live simply. The Mbuti people who live in the rain forests of central Africa gather berries. They catch fish and crabs from the Ituri River. The Mbuti sometimes collect honey from beehives. Hunters may set traps to catch birds in the trees.

Many rain forest people traditionally did not wear much clothing. The hot, humid climate of tropical rain forests kept them warm. Today, the Yagua people in the Amazon wear skirts made from palm tree fibers. They even dye the skirts red. How? They get the color from the annatto seed.

Decorations

Some rain forest peoples decorate their bodies with paint, ink, or jewelry. They may use painted designs or tattoos to cover their skin. The decorations sometimes mean something special to the tribe. Native groups also make jewelry from rain forest materials. These include nuts, shells, feathers, and stones.

These masks from the Amazon Rain Forest are made from materials found nearby, including feathers and nuts.

HOUSING

Traditional rain forest homes are built with natural materials. In South America's rain forests, most are made from logs covered with grass or palm leaves. The Mongo people in the Democratic Republic of the Congo make roofs from palm leaves. They also use sticks, mud, and anthill clay for some buildings.

Some rain forest peoples live in houses high above the ground. This helps protect them from animals, mosquitoes, and floods. This is how the Korowai people live in New Guinea's rain forests. Their tree-house homes can be as high as 150 feet (45 m) above the ground!

Rain forest peoples often sleep in hammocks made of palm tree fibers or cotton. Many native rain forest peoples live in shared houses. The Amazon's Yanomami and Enawene Nawe peoples live in group homes.

This hut on stilts is one type of home found in rain forests. →

MODERN LIVING

Millions of people live in rain forests today. Many live in modern towns and cities. Others live in small villages like their ancestors did long ago.

Cities in the rain forest are just like other cities. They have stores, schools, businesses, and restaurants. Manaus, Brazil, is a city in the Amazon Rain Forest. About 2 million people live there. Setulang is a village in Indonesia. Less than 1,000 people live there. People farm the land using methods from the past. They also hunt in the forests.

BRAZIL

Colorful houses make up this crowded section of a Brazilian city, which is bordered by tropical forest.

Many different **industries** depend on the rain forest. Rubber is one example. It is made with a liquid that comes from trees growing in the rain forests of South America and Asia. Rubber can be used to make shoe soles and many other items.

Many rain forest **residents** work in the timber, or wood, industry. People cut down trees to make furniture and paper. Trees are also cleared for mining. Oil and metals like copper, tin, iron, and gold are found in rain forests.

Farming and cattle ranching are common industries in many rain forest areas. **Ecotourism** is also becoming popular. Visitors from all over the world come to see the rain forests' plants, animals, and landscapes.

There are many ways to get around in the rain forest. The most common way is by walking. Rivers are like nature's highways for rain forest residents. Many people travel in canoes. Deep canoes called dugouts usually hold one to six passengers. Other boats, like a *rápido*, have powerful engines. These let people travel fast.

People travel through rain forests on boats like this one.

HEALTH CARE

Modern technology is not part of all rain forest peoples' daily lives. In large areas of the Amazon Rain Forest and other rain forests around the world, there isn't much cell phone coverage. That means people can't use cell phones!

Still, some places in the rain forest are using technology more. Rain forest homes may have electricity. They might have small freezers to keep meat and other goods fresh. Some even have computers and telephones to communicate with people in other places around the world.

Health care uses advanced technology in some rain forest communities. Nurses may see patients in the rain forest. They then use computers to talk to doctors in different cities. A patient can "see" an expert far away for help with illnesses or other health problems.

Tropical forests surround a countryside school in Nigeria.

SCHOOL IN THE RAIN FOREST

Brazil doesn't have enough teachers to reach all the kids spread throughout its rain forest. Distance learning, or learning from far away by using a computer, is helping. Several new schools have classrooms and computers that can connect to the Internet. Students travel by boat to one of these schools. They stay at school for 10 to 15 days each month. Teachers who are hundreds of miles away in the city of Manaus then give the students lessons over the Internet.

PROBLEMS IN THE RAIN FOREST

Rain forests are important. They are where many plants and animals make their homes. Over 2 billion people depend on the world's forests for their food and shelter.

Rain forests around the world are shrinking. This loss of forests is called deforestation. Farmers cut down trees so they can plant **crops**. They also clear forest land to raise cattle or sheep. Logging, or cutting down trees to sell their wood, causes deforestation. Rain forests are also being cut down to make space for new cities. If deforestation continues, some say Earth's rain forests could be completely gone in 100 years.

Rain forests create 40 percent of Earth's oxygen. When trees make oxygen, they also take in a gas called carbon dioxide. With fewer trees, more carbon dioxide stays in the air. This causes **climate change** and hurts the planet.

Rain forests on the island of Borneo have been cleared to make room for a farm.

Endangered Animals

When forests shrink, animals lose their homes. This is called habitat loss. There are many **endangered** species in rain forests. Some of these animals lose their homes because of deforestation, but another problem for them is poaching, meaning illegally killing or capturing animals. Poachers trap rain forest animals to sell as pets or kill them for their body parts.

HOPE FOR THE FUTURE

Countries around the world are trying to stop deforestation. People plant trees in forests that have been hurt. Some nations, like Malaysia, are creating new parks to protect rain forests and the plants and animals that live there.

What can you do to help? Plant trees! If more trees are planted on Earth, more oxygen will be made. Another helpful thing to do is to become an ecotourist. Ecotourism is a great way to learn about nature in an area. It also shows others you are interested in protecting Earth.

People will probably live in rain forests for a long time. With teamwork, the future of the rain forests could be amazing. However, it can only happen if people like you do your part.

RED COLOBUS MONKEY

MAGOMBERA NATURE RESERVE

Magombera Forest is in the African country of Tanzania. It is home to unique, or special, animals like the Udzungwa red colobus monkey and the greater bush baby. In January 2019, a new nature reserve was created to protect this rain forest. Many endangered plants and animals will now have a safe place to live.

This young man is taking care of young trees near the Amazon Rain Forest of Brazil.

WORLD'S FIVE LARGEST RAIN FORESTS

RAIN FOREST	LOCATION	SIZE (SQUARE MILES)
Amazon Rain Forest	South America	2,100,000
Congolese Rain Forest	Africa	687,260
New Guinea Rain Forest	Oceania	111,200
Valdivian Temperate Rain Forest	South America	95,800
Heart of Borneo	Asia	84,950

AMAZON RAIN FOREST

GLOSSARY

climate change: The process by which Earth's normal weather patterns are shifting due to a rise in global temperatures.

crop: A food, such as a vegetable, that is grown by farmers.

ecotourism: Tourism, or travel, focused on nature, especially to observe wildlife in a way that doesn't harm the environment.

endangered: At risk of dying out.

equator: An imaginary circle around Earth that is halfway between the North Pole and the South Pole.

evergreen: A tree that is green all year round.

generation: A group of people one step in time from another group, such as in a family; children are the next generation after their parents.

industry: A group of money-making organizations or businesses that do the same kind of work, such as hotels or car factories.

rare: Hard to find; few in number.

resident: A person living in a particular place.

species: A group of living things that are all the same kind of plant or animal.

temperate: Having a climate that is normally mild without extremely hot or cold temperatures.

tribe: A group of native people who live together and share many beliefs and practices.

tropical: Warm and humid with lots of sunshine.

INDEX

WEBSITES

Due to the changing nature of Internet links, PowerKids Press has developed an online list of websites related to the subject of this book. This site is updated regularly. Please use this link to access the list:
www.powerkidslinks.com/hdylt/rainforest